Plant Leaves

1.

Plant Leaves

Prints by

KIPTON KUMLER

Introduction by

HILTON KRAMER

David R. Godine · Publisher

David R. Godine, *Publisher*
306 Dartmouth Street • Boston, Massachusetts 02116

Copyright © 1978 by Kipton Kumler
Introduction copyright © 1977 by Hilton Kramer

No part of this book may be used or reproduced in any manner whatsoever without written permission except in the case of brief quotations embodied in critical articles and reviews.

ISBN 0-87923-258-7 • LC 78-58508

Printed in the United States of America

Introduction

HILTON KRAMER

The photographer's eye dwells in two worlds. Each is a universe bounded by light and space and the objects they enclose, and each depends on the other for its photographic existence. One is the world 'out there,' which, whether a few inches distant in the same room or a long journey to a faraway place, is where the photographer discovers his subjects. The other is the world of the darkroom where, in a sense, he rediscovers them in the process of giving them a permanent visual form. Great differences — in scale, pace, dynamics, concentration, materials, and technique — separate these worlds, yet one of the great traditions of photography (it is not the only tradition, of course) rests on their organic alignment and unification. For photographs of this tradition are not something merely 'taken'; they are something made. And the making of them is no less integral to their conception than the taking of them.

Kipton Kumler is a photographer very consciously of this persuasion — that is, an artist who has elected to work within the confines of an established tradition. His entry into photography came through a course of study with Minor White at M. I. T. He then studied with Paul Caponigro. These are names that in themselves define the tradition in which Kumler works, and through them we are made to feel the presence of the elders who created it — Stieglitz, Strand, Weston, and Adams. Theirs is, first of all, a fine art tradition, with an exalted sense of craft and painstakingly esthetic in its fundamental outlook. This outlook governs both its choice of subjects and its attitude toward the medium. The subjects are, more often than not, drawn from nature, with which the photographers of this tradition seek to establish a special relation. Toward both nature and the medium, indeed, the attitude is reverential, at times even mystical. There is a spiritual cast to it. There is a distinct appetite for purity and the absolute.

To undertake to work in such a tradition in the hope of achieving something distinctive involves obvious hazards. One is obliged to master difficult disciplines, and one knows at the outset that the result will be judged by standards almost impossible to meet. All access to shock, surprise, novelty, to the kind of disrupting cultural assault that has long been associated with contemporary taste in the arts, is foreclosed. The risks of merely repeating the past are great. Achievement, when it comes, is bound to be incremental.

The challenges are not those of the stunning, all-eclipsing 'breakthrough.'

Kumler has faced these hazards with a remarkable confidence and equanimity. Technically he has equipped himself to do — and do flawlessly — whatever the standards of this tradition require, but he also has a clear grasp of the place of technical excellence in the art he practices — of its role as a condition and coefficient of vision rather than its object. And it is, above all, in the realm of vision — in the quality of the images he has wrested from nature and the world 'out there' — that Kumler has won his place in the tradition in which he works. He has the requisite eye for this arduous labor, an eye that is at once contemplative and analytical — an eye that is open and available to the lyrical occasions his subject naturally offers, yet disciplined to insist upon, and get, the kind of preternatural clarity and formal definition that is so elusive in nature and so essential in art. For 'nature' in photography of this persuasion is not the nature of nature; it is the 'nature' of art. It too is something 'made' — made in the act of observation and the mystery of response. It is something fixed and finite, something that the nature of nature can never be. It represents neither an improvement upon nor a diminution of actual nature, but something separate and parallel — an esthetic correlative that gives form to an emotion.

In the suite of greenhouse pictures that have been Kumler's special concern in recent years, we see the impulses of this art in a particularly concentrated form. These are photographs of plant life and of the light that nourishes that life. And these too are photographs in a tradition, for no photographer after Weston could fail to perceive the affinities that exist between the structures of plant forms and the forms of art. Kumler assumes the perception of these affinities as the *donnée*, not as the object of discovery, in this project, and carries on from there in the environment of art.

For nature in the greenhouse is no longer the nature of nature either. It is nature on its way to becoming a form of artifice — it is nature placed under the governance of applied intelligence and control. In the greenhouse, the fundamental rampancy of nature is disallowed. All growth is shaped and directed and disciplined. To the extent that the biological impulse allows, plant life in the greenhouse is turned into a form of still life. It continues to burgeon, in fact it flourishes — this is the very point of the control — but only within a restricted course. The greenhouse is, in effect, a world of ideal forms. It is nature removed from the contingencies of nature, and made into a form of culture.

The photographer who ventures into this world knows very well that he is engaging in a labor of collaboration. His purpose cannot be to summarize it, for even the artifice of the greenhouse is beyond summary; in that sense, at least, nature in the greenhouse remains part of the nature of nature. His perceptions are bound to be fragmentary, and are accepted as such, and it is out of these fragmentary perceptions that he creates his forms. The result is, in a way, a kind of visual allegory on the nature of art and its relation to the nature of nature. And one of the things this allegory tells us is that any object of nature that is made into an object of contemplation has the power to affect us as an object of art. The medium becomes a mode of perceiving both itself and the thing outside itself.

Out of his own fragmentary perceptions in the greenhouse, Kumler has given us a kind of photographic chamber music. His pictures have the delicacy of chamber music, the intimacy too, and the sustained emotion that comes from being so close-up to the details of articulation. In these pictures, elegance lives on easy terms with austerity. The eye for form is flawless, yet a continual delight. There is a remarkable warmth in them, an affectionate identification, yet an artistic detachment that keeps these images from being merely 'personal.' They are indeed something made, and not only observed, and in the quality of their making Kipton Kumler has succeeded in renewing the tradition in which he now takes a distinctive place.

The Prints

2.

3.

4.

5.

6.

7.

8.

9.

10.

12.

13.

14.

15.

16.

17.

18.

19.

20.

21.

22.

23.

24.

25.

Editorial Note

Prints numbered 1, 3, 4, 7, 10, 13, 20, 21, 22, 23, together with the Introduction, originally appeared as a bound portfolio of platinum prints in an edition limited to fifty numbered portfolios with an additional seven lettered portfolios hors commerce.

PLANT LEAVES was designed by Katy Homans and set by Michael & Winifred Bixler in Monotype Centaur. The paper is Mohawk Superfine, an entirely acid free sheet. The plates have been printed in two color, fine line duotone by The Meriden Gravure Company from separations made by Richard Benson.